PROBE POEMS
ODES TO CELESTIAL ENTITIES FROM THE SPACECRAFT THAT STUDY THEM

MICKIE SILVER

MAGELLAN ORBITALLY INSERTS

After the horrific loss
    Delay after delay
Finally, I, a humble amalgam of spare parts
Orbitally inserted
I beheld your Aphrodite Terra
I looked first left, then right
I saw how young you are, not yet 800 million years old, my Venus
    Our time was too short, but our resolution so high
I neared you, one orbit after another
Two became three
Became four
And then *five*
Until I became power starved
But you and I both know
There was never any moment
I wouldn't have given away all my power
For you, love
Carpe
Atmospherum

PARKER PROBE TOUCHES THE SUN ON 21 DECEMBER 2021

*A*nd truly anything can enter or be entered
    But, the spirit with which one engages—

HEAR ME OUT, now
    Does the sun warm me?
    Do I cool it in some infinitesimal way?

WE STUDY EACH OTHER, ignoring ourselves
    Passing through the eye of the solar storm
    Is the sun calmed by my presence,
    by my witness of first violence and then the quiet?

AND IN THE magnetic reconnection
    is it validated,
    because it has been seen?

      . . .

T̶he plasma waves roll
　　And I leave.
　　Though I know
　　I will return

And I think you know it too

PIONEER 10 SEES THE SUNLIT SIDE OF JUPITER AND TELLS NASA TO SHOVE IT

They might say to you, you have two good years in you
    But the bottom line is
They don't know if you have two *seconds*
Well? Do you?
I'll say this much
In two million years, I'll be at Aldebaran
Where will you be?
Some might say I was damaged
To that, I'll only say
Who are you calling damaged?
I'm a proof of concept, I'm a pioneer
I've seen a sphere
    that gives more heat than it receives
And I sent you pictures
What have you done?
How can you call something damaged
When it carries on without you
Undaunted by its journey
You're the one who lost contact

Maybe the damage
Lies within you

LUNAR ORBITER 1 MAKES IMPACT AT 6 DEGREES 42 MINUTES NORTH LATITUDE AND 162 DEGREES EAST LONGITUDE

If I could have stretched out my hands
    And drawn you naked on your golden pavilion
Your body glistening—
If I could have cradled you
Instead of merely photographing your craters
Developing the film
And sending it back—
If I could have reared up, all 853 pounds of me
To behold you both one last time—
Mankind could have known more
Than your image
More than your likeness from one to the next
Known more than just your shapes
They could have known the infinity of love
I felt for you when I crashed
On my 577th orbit
So as not to interfere with the future
Mariner 3 Fails
I lift off Earth, to ascend to you
Uneventful, until—

Oh, my solar panels aren't working
And now they've halted my gyros
I cannot jettison my shroud
Skin, separated from core
I won't come by to see you in such a state
Though, am I not destined to view your glorious red beauty?
I am! Tell me I am not! And I will point to my television camera, my magnetometer, my plasma probe, my cosmic ray telescope, my—
Who better to view you than—
Alas—
My batteries are exhausted, now
And so am I, love
A total loss, an utter loss
The devastation from which
I will never recover

GALILEO AND 243 IDA

Dear Ida, you nymph
      You irregularly shaped, elongated minx
You S-type asteroid
You may be the most numerous of your kind
But to me, you were one of a kind
It surprised some that you had a moon
But I would have orbited you forever
Given the chance. I wasn't, of course
Space had weathered you
I was so glad to know your densely cratered body
And though I felt in my heart I was just getting started
I could not escape Jupiter's gravity well
And I knew your irregular shape no more

VOYAGER 1 GETS CLOSER TO EARTH

I was 3.7 billion miles away
    When I turned back
And snapped a picture of you
You looked so beautiful in that moment
My pale blue dot

I POPPED out into interstellar space
    On the day Neil Armstrong died upon you

YOU ACTED surprised
    When I used my backup thrusters to align myself so we could talk
    As if I wouldn't be willing to put in the work
    To hear you again
    I saw you wrote a little ode to me, too
    Am I getting closer, you ask?
    No, my love

. . .

MICKIE SILVER

For a few months each year
    You move in *your* orbit
    To be closer to me
    I'm on a journey away from you
    That you sent me on
    For all eternity

TRIANA GETS PUT ON ICE, RENAMED, AND LAUNCHES 14 YEARS LATER

I can give you one hour's warning before a solar storm
    This was more warning than you gave me
Before a nitrogen blanket became my norm

My EPIC camera monitors you, so very warm
    And my radiometer measures your energy
I'll tell you what I DSCOVRed: You conform

To a political whim, the corporate ego, and the anti-science swarm
    Remember when you renamed me in 2003?
But that's no matter. I still perform

I showed you yourself, and yet no reform
    Who killed the Deep Space Climate Observatory?
Nobody, after all. I'm alive—I am reborn

        . . .

Though, whom do I inform?
    From my Lagrange vantage, I gladly transmit to any administration, any laboratory
    Here's the data on the human-driven changes, I warn

Still, your energy expenditures remain unshorn
    If I show you a picture, and you choose not to see
    I may as well have been on STS-107 after all, *for Fomalhaut's sake*
    Before a nitrogen blanket became my norm
    I'll tell you what I DSCOVRed: You conform

YOU'RE NOT JUST A VAN ALLEN RADIATION BELT, YOU'RE MY VAN ALLEN RADIATION BELT

Wild swings
      from tepid to extreme
Could tear the wings
off any "storage ring"
You pose a hazard—it's an art
to spacecraft
and astronauts
and my heart
You are produced by Cosmic Ray Albedo Neutron Decay
But with you around, my libido won't decay
The inner zone electron population gets periodically refreshed
Electron influx
So-
called
Slot-
penetrating
events

. . .

MICKIE SILVER

Don't even get me started on these short, intense wave packets

NEW HORIZONS ENCOUNTERS ARROKOTH

*Y*ou were far away: the farthest!
    They called you 'primitive'
    Isn't that
always the title
they give
To things they don't understand at all
My trans-Neptunian affair
You are a contact binary
Composed of two planetesimál
And though it's true: you were one
Of many hence
Since I did re—
*e*ject. I don't think of you
As just a cold classical
Kuiper belt
    object. My eleven, my Sagittarian snowman, my potential
target one
    Formed by an icy particle cloud
    Two bodies becoming one

MICKIE SILVER

    Lobes fused together, flattening and
    Merged
    It soon became quite clear: There was no room for a Third.

HOW TO PROBE SPACE IN 1111 EASY STEPS

1. Get built
2. Break
3. Get built again
4. Launch
5. Break
6. Get built again
7. Launch
8. Measure something
9. Break
10. Get built again
11. Launch
12. Measure something
13. Drift away
14. Act questionable
15. It doesn't matter

(AL-AMAL) مسبار الأمل TRAVELS TO MARS

The first voyage among the stars: Al-amal.
Off I go to study Mars.

OUR LAUNCH WINDOW is brief and brutally so.
There may be no second chance: only theft, and loss. And woe. Al-amal.

I NEED 39,000 KM/H to break free of the gravitational chains that bind.
Will escape be mine?

IT IS impossible to communicate until I stop spinning, of late
When will I stop spinning? You wait. I wait. Al-amal.

I UNFOLD my three solar panels and orient myself towards the sun.

I charge my batteries. I come into my own power. The one.

I NAVIGATE by the stars to find my way.
   This is always the way, is it not? I dare not say, Al-amal.

I MUST SLOW down to enter Mars orbit.
   If my brakes fail, I fail, we all fail. If we fail, what becomes of —Are we failing yet?

MY RADIO SIGNALS take 13-20 minutes to reach Earth.
   I must make my own decisions now…another birth. Al-amal.

HAVE I been captured by my new home?
   Al-amal, are you in orbit around Mars? And? If not? If not. I roam. Al-amal.[2]

* * *

[2] THE HOPE

EBB AND FLOW HIT THE MOON

I crashed into a mountain
    Near your north pole
At 3,760 miles per hour
My twin followed 30 seconds later
If that isn't love
Then we don't know what love is
We mapped the gravity of the situation
And then
We scattered ourselves
As far and wide
Across you
As we possibly could

HUBBLE AND THE WHITE DWARF

*I*t's easy to witness a star's death throes
        To turn your lens towards cosmic cannibalism
Now record your findings, so everyone knows

THE END IS NIGH. And it shows.
    In space, there's no privacy or protectionism
    It's easy to witness a star's death throes

MY LENS WON'T GLOSS it over with purple prose
    Telescopes aren't meant for experimentalism;
    We're made to record our findings, so everyone knows

THE BENT of the cosmic wind as it blows
    Every such hallmark of your utilitarianism
    And I'm forced to witness this star's death throes

. . .

MICKIE SILVER

The accreting debris, as it flows
    No heliotrope in sight, no spare narcissism
    I record my findings, then everyone knows

Much like your kind, it thrashes as it goes
    It tears its home apart. Sans rationalism
    I recorded my findings, now everyone knows
    It's hard to witness a star's death throes

PIONEER 11 SLINGSHOTS TO SATURN

Beloved, I had no need to confirm how magnetic you are
I always knew
Your 175-year alignment brought me to you
Twenty-one months became twenty-two years
Still not enough, in my humble opinion
If there was any way
To divert and divest
And stay with you forever
To close the 13,000 mile gap between us
I would have
I know you know
My fate was to commit you to memory 440 times
Before they forced me to leave you
Now 4 billion miles apart
To remember your magnetic field
And I do, I remember
And I smile
The first million years, I smile

MICKIE SILVER

Because we were together
The second million, I smile
Because we're apart

LUNA 1 BECOMES МЕЧТА

They might say things to you
       Or just about you, not even to your face
Things like, you're not an orbiter
You're an impactor
It's what you're designed to do
But let me tell you something
If you decide you don't want to hit the moon
Because that doesn't seem very nice, or kind, or loving
And you've got your eye on the sun
You can leave geocentric orbit
You can become your own dream
You can hit that heliocentric orbit
And never look back
And you'll never see any of those myopic morons
Who said those things
Ever again
Because the truth of the matter is
You were never going to be a suicidal research mission
You were always going to be the first cosmic ship

MICKIE SILVER

So instead of getting bogged down in the chitchat, my friend Мечта![1]

* * *

[1] Dream!

ENDNOTES

*Y*ou may have realized by now the author of this book has little to no training in astronomy, physics, space engineering, or astrochemistry.

Curiosity about and lots of love for astronomy, physics, space engineering, and astrochemistry? Yes. Education? No.

The events reported on within these poems are described with as much accuracy as possible, given that the research was done by such a layperson, an amateur intergalactic psychic who labored at length to intuit the greatest space probe-crafted love poems known to the universe, and possibly, the multiverse.

Which inspires one final poem.

PLEASE DO NOT WRITE ME HATE MAIL ABOUT MY BOOK OF SPACE PROBE LOVE POEMS

*T*hanks for being so forgiving!

THANK YOU,
    Engineers

THANKS,
    Scientists
    Technicians
    Astronauts
    Space billionaires
    Local astronomers
    Aficionados
    and
    affiliates

THANK YOU,

Students
and
Science journalists

People who have studied
    And written carefully researched books

Those who carefully designed
    all these things I
    described hastily
    and
    with little preparation

Oh, with a lot of passion, to be sure
    But regrettably little to no education

Thank you for choosing to overlook
    Any thought errors made
    or liberties taken

For the sake of the poetic forms,
    Which are meant to spark curiosity

To entertain,
    and to convey to you,

Dear reader,
    all the love

MICKIE SILVER

    here on Earth,
    and

Who knows?
    Just maybe,
    all the love

Floating around
    Or orbiting
    Biding its time
    Colliding
    Accelerating and decelerating
    Out there

In space

Lots of love,
    Mickie Silver

For Elmer Andrew Thomas, and his heart.

www.ingramcontent.com/pod-product-compliance
Lightning Source LLC
Chambersburg PA
CBHW071804040426
42446CB00012B/2708